COTTON and SILK

Jacqueline Dineen

ENSLOW PUBLISHERS, INC.

The marvellous stories of how cotton is produced from the cotton plant, and silk from caterpillars' cocoons, ⬛ made into cloth.

Contents

*The picture above shows a cotton 'boll'
– the seed pod containing the fibre from
which cotton is made.
[cover] The cover picture shows a
cotton harvester at work in the southern
United States.
[title page] A worker gathering
silkworm cocoons on a farm in Sichuan
Province, China.
[1–25] All other pictures are identified
by number in the text.*

This series was developed for a worldwide market.

First American Edition, 1988
© Copyright 1985 Young Library Ltd
All rights reserved.
No part of this book may be reproduced by any means
without the written permission of the publisher.

Printed in the United States of America

10 9 8 7 6 5 4 3 2 1

Library of Congress Cataloging-in-Publication Data

Dineen, Jacqueline.
    Cotton and silk.
    (The world's harvest)
    Summary: Describes how cotton and silk
are naturally produced and how they are
prepared and made into cloth.
    Includes index.
    1. Cotton--Juvenile literature. 2. Silk
--Juvenile literature. [1. Cotton. 2. Silk]
I. Title. II. Series: Dineen, Jacqueline.
World's harvest.
TS1576.D56 1988   677'.21   88-1180
ISBN 0-89490-213-X

Introduction

If you look around your house, you will probably find a lot of things made of cotton. Cotton is used for clothes, bed linen, curtains, furnishings, and many other things. About half of all the textiles produced in the world are made of cotton.

Silk is popular too, but in a different way. It is expensive and luxurious, and is used to make elegant clothes and sumptuous furnishings. Both cotton and silk are completely natural fabrics. [1]

First, I tell you about cotton. Cotton is a plant, which grows in warm climates. In chapter 1 I tell you how countries like the United States have huge cotton plantations where the work is done by machines. In other parts of the world, like Africa and India, there are small cotton farms where the work is done by hand. I tell you about both these types of farming, and how the cotton is planted, cultivated, harvested and sold. Picture [1] shows bales of raw cotton being unloaded at the docks.

In chapter 2 I explain how the fibre is separated from the seeds. Then we go to a spinning mill to see how the cotton is turned from a mass of fluffy fibres into long strands that can be woven into cloth. I explain how the cotton is woven, dyed, printed and finished.

Silk is produced in a completely different way. It comes from silkworms which are reared on silk farms. In chapter 3 I tell you how the worms are hatched, fed, and then given a place to spin their silken cocoons.

In chapter 4 we see how the cocoons are treated so that the silk can be unravelled and spun. In some parts of the world this is done by hand, and in other places it is done by machine. I tell you about both methods. Then I describe how the silk is woven into beautiful and delicate fabrics.

In the last chapter, I tell you about some of the ways cotton and silk are used. You may be surprised by some of these! I hope you like my book, and enjoy learning the curious and interesting stories of these very different fabrics.

A silkworm is not really a worm. It is the name given to the caterpillar of the silk moth. Its life cycle is described in chapter 3.

1 · Cotton farming

Like wool, cotton is a natural fibre which has been known to people for thousands of years. Unlike wool, it does not come from animals. Cotton grows on the cotton plant, and is a fluffy mass of fibres that surrounds the seeds, which you see in picture [2].

The cotton plant likes a warm climate with plenty of sun. It is grown in eighty-five countries which lie between latitudes 45° north and 30° south. Look at a world map and pick out the cotton-growing belt. The main producers are the United States of America, Russia and China. Other large producers are

[2]

India, Pakistan, South America, Africa, Turkey, Greece, and the Sudan. Cotton farming is a main source of livelihood to 125 million people in developing countries, and to many others in richer countries. So where would we all be without it?

Cotton is popular because it is so cool and fresh to wear. No man-made fibre can compare with the feel of real cotton, especially in the heat. Cotton fibres allow the skin to 'breathe' and they also absorb moisture and allow it to evaporate. Cotton clothes keep you cool and comfortable in hot, humid weather for far longer than clothes made from artificial fibres.

The cotton plant, *Gossypium*, looks a bit like a blackcurrant bush. It is grown as an 'annual'—this means that new plants have to be grown every year. You can see a row of the young seedlings in picture [3].

The planting season depends on the part of the world the cotton is grown in. In the northern hemisphere it is planted in the spring. In the southern hemisphere it is planted in December.

Methods of growing cotton vary from country to country. This is partly to do with the climate, and partly to do with the size of the farm. In the United States, for example, much of the work on the huge cotton plantations is done by machine, like the sprayer in picture [4]. Machines plough up the soil and prepare it for planting. Other machines sow the seeds. Aircraft fly over, spreading pesticides and fertilizers on the crop.

On small farms in India and Africa, most of the work is done by the farmer and his family.

Man-made fibres are fibres which are not produced from natural materials such as wool, cotton, or silk, but are manufactured from chemicals.

[3]

They use oxen or buffalo to pull the plough, [4]
and the seeds are sown by hand.

The cotton plant does not like the cold, so it
must be planted, grown, and harvested, before
there is any sign of winter frost. In some areas,
such as southern Europe and the United
States, this is a race against time. Cotton
farmers choose the type of cotton seed that is
most suitable for their climate. Some types will
grow in desert regions so long as the land is
well irrigated. Other types will grow in the
more humid areas near the equator.

About eight weeks after the seeds have been
sown, creamy yellow flowers appear on the

[5]

bushes, like the one in picture [5]. The flower falls after a single day, leaving a small pod. This pod grows into a cotton 'boll'. Over the next two months the boll swells until it is the size of a hen's egg. You can see a picture of the boll in this state on page 2. When it is ripe, it bursts open to show the cotton fibre inside. The fibre is a fluffy mass surrounding the cotton seeds. Each boll contains about thirty seeds.

When the boll has burst, the fibres dry in the sun and the cotton is ready for harvesting.

On small farms, the bolls are picked by hand. In some ways this is better than harvesting by machine, because the bolls dry and ripen at slightly different times. The pickers can select the ripe bolls and leave the others to be harvested later.

The farmer and his family work hard to harvest the cotton. Everyone helps, and sometimes extra pickers are employed to get the harvest in. In picture [6] the pickers move along the rows of bushes, choosing the best bolls and putting them in a bag or basket slung over their shoulders. They are careful to pick only the boll and not the leaves or the outer boll shell.

The picked cotton, with its fluffy fibres and seeds, is called seed-cotton. The seed-cotton is spread out to dry before being packed into sacks and sold at the local cotton markets.

On the large plantations, it would take too long to pick the cotton by hand. Mechanical pickers, like the one shown on the cover of this book, can get the harvest in very quickly. A machine can pick as much cotton in one hour as a person can pick in a day. However, there are disadvantages too. The machine doesn't know which are the best bolls to pick, so it picks them all. It also picks the leaves and boll cases, which all get mixed up with the seed-cotton. The cotton has to be carefully sorted later to get rid of the 'trash', as this extra material is called. Sometimes the bushes are 'defoliated' before harvesting so that there is not so much 'trash' to pick up. A chemical is put on to the bushes to make the leaves fall off, leaving the bolls behind.

The raw cotton is sold just as it is. Later the seeds and the trash will be removed from the fibre before it is spun into yarn. [6]

2 · *Making cotton*

[7]

[8]

When the seed cotton has been harvested and dried, it is taken to 'ginneries'. There the fibre is separated from the seed.

In developing countries, where a farm produces a fairly small amount of cotton, the farmer packs up his crop and takes it to a local seed-cotton market. Picture [7] shows a seed-cotton market in Africa. A buyer from the national cotton marketing board is inspecting the cotton, and he will pay the farmer according to its quality. The buyer then arranges for the seed-cotton to be taken to one of the national ginneries.

At the ginnery, the fibre is separated from the seeds by a ginning machine. You can see a ginning machine in picture [8]. The seed

cotton is fed into the teeth of a row of circular
saws. These saws pull the fibres through
narrow slots. The seeds are too big to go
through, and are left behind.

The cotton seed is not wasted. Animal feed is
made from the husks and kernels. The oil is
used in food production.

Now the cotton fibre is ready for the mill. It
will be shipped to mills all over the world, to
be spun into cotton and woven into cotton
fabric. You would not get much into the ship's
hold if you left the raw cotton loose and fluffy,
so it is pressed tightly into bales. Picture [9]
shows the compressed bale being secured with
metal bands.

Before it leaves the ginnery, samples are
taken from the bales. These are used to show
the buyer how clean the cotton is, and how

long and strong the fibres are.

In large cotton-producing countries like the United States, the cotton farmer sells his crop direct to a ginnery. When the ginnery has turned the seed-cotton into cotton fibre, it is sold to a cotton merchant. The cotton merchant sells the cotton to the world's markets. Some is sold to mills in the cotton-producing country, and the rest is exported. The United States is the largest exporter of cotton in the world. The other main producers—Russia and China—keep a good deal for their own use.

Long before the cotton has even been harvested, factories have already ordered huge amounts from the cotton merchants. These factories are in countries which specialize in turning cotton into cloth. For example, Hong Kong and Taiwan turn the cotton into textiles, then export the textiles to other countries. So when the bales are ready, the merchant arranges for them to be shipped to the countries which have ordered them.

The countries which buy cotton

However, there are also cotton merchants in countries which produce no cotton. In cotton-importing countries like Britain, France, and Germany, merchants import the cotton and then sell it to cotton mills. The cotton is shipped into one of the 'cotton exchange' ports such as Liverpool in England, Le Havre in France, or Kobe in Japan.

At the port the merchants 'class' the cotton so that a price can be set. They take samples

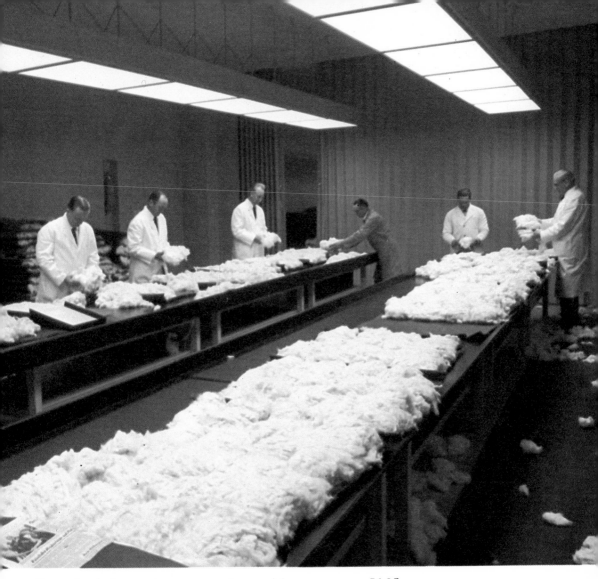

from the bales and compare them with
standard samples which they keep for this
purpose. You can see this being done in picture
[10]. Buyers want to know how long the fibres
are, how clean and white the cotton is, and
whether it is coarse or fine, before they decide
which bales to buy.

[10]

The bales of raw cotton are sent to the spinning mills. Here the cotton will be turned into a continuous length, called yarn. The yarn will then be sent to a weaving mill to be turned into fabric.

Each bale, when it arrives at the mill, is a mass of tangled fibres pressed tightly together. Bits of leaves and twigs from the plants are still caught up in them. First the bales are fed into opening and cleaning machines like the one you see in picture [11]. It has opened the bales and separated the cotton into small tufts.

Some types of cotton have mainly long fibres, others have short or medium-length

[11]

[12]

fibres. The longer the fibres are, the finer the
yarn that can be made from them. Most yarn
is made from a blend of different types of
cotton. So cotton from different bales is fed into
the machine and blended together.

The tufts are collected into a fluffy mat
called a 'lap' and fed into a carding machine.
Carding untangles and cleans the fibres. The
cotton is pulled over wire points to separate it
into individual fibres, and remove bits and
pieces such as leaves and twigs.

The cotton comes out of the carding
machine as a filmy web of fibres which you
can see in picture [12]. Fibre intended for
high-quality cotton yarn is combed to remove
some of the short fibres. The cotton is now
ready to be spun.

[13]

Spinning is the process of drawing out the fibres into a continuous strand and twisting them together into yarn. First, the fibres in the web are brought together to form a long rope called a 'sliver'. The sliver is about 2.5 centimetres thick and the fibres are not straight. The sliver has to be 'drawn out' to make a longer, thinner strand. Several slivers are drawn out together, to blend the fibres again. The slivers pass through rollers which are travelling at different speeds. These pull the fibres out into a fine sliver called a 'roving'. It is this roving which is spun into yarn.

The yarn is usually spun on ring spinning frames like the one in picture [13]. The roving is drawn out to the right thickness, twisted into yarn around a spindle, and wound on to a ring bobbin. A more modern method is called 'rotor spinning'. Fibres are collected on the inside surface of a high-speed rotor. As the machine rotates, it twists the fibres into yarn.

At the weaving mill

Next, the yarn is either woven or knitted into fabric. In weaving, lengths of yarn are criss-crossed together. The vertical threads (the 'warp') are set up on a weaving machine called a loom. The horizontal threads (the 'weft') are passed under and over the vertical threads, again and again, from top to bottom, so that the intertwined threads form a sort of mat. Picture [14] shows one of the large mechanical looms on which cloth is woven.

Patterned cloth can be made by using yarns which have been dyed in different colours. An additional way of making a pattern is to alter the weave. You could go under one warp thread and over three, for example, or under two and over one.

Weaving is one way of making cloth. The other way is by knitting. Cotton can be knitted by hand, but is more usually knitted on machines. It makes very lightweight material for underwear and many types of fashionable clothing.

If the yarn has not been dyed before weaving, the cotton which comes off the loom is known as 'grey' cloth. This means that it is a

[14]

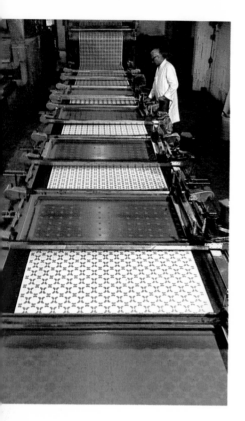

[15]

dirty white colour and needs to be thoroughly cleaned and bleached to make it pure white. Some cotton is left white, and some is dyed in a variety of plain colours. Sometimes a pattern is printed on to the cloth in a roller printing machine. The pattern is engraved on to rollers. The machine uses colour pastes made of dye and a thickener to print the pattern on to the cloth.

Another way of printing patterns on cloth is by flat screen printing. Each colour needs a separate screen. The pattern is set up by masking out parts of the screen, and colour passes through the unmasked areas on to the cloth. Picture [15] shows flat screen printing machines. Rotary screen printing works on the same principle, but is much faster because it is a continuous process.

All that now has to be done is to 'finish' the cotton. Finishing may just mean ironing the cloth to give it a smoother surface. But it usually means much more. If the fabric is going to be used for a special purpose, such as keeping out the rain, it is given a particular finish at this stage. Fabric for raincoats, for example, is given a water-repellent finish to make it waterproof. Material for industrial clothing may be given a fire-repellent finish. 'Easy-care' finishes improve the appearance of cotton fabrics. Material for clothing and household items like sheets are often given an 'easy-care' finish.

Before the cotton fabrics leave the mills, they are inspected carefully for weaving and dyeing faults. Then they are ready to be sent to clothing factories, and finally to the shops.

3 · *Silk farming*

Silk is an expensive and luxurious material. Even the word 'silk' sounds romantic. So it is strange to think that such a precious fabric starts life in a very humble way. It is merely the cocoon which a certain caterpillar spins around itself when it is ready to change into a moth. Picture [16] shows what these caterpillars look like.

All caterpillars spin a silk cocoon when they are ready to change into moths or butterflies. The cocoon hardens into a pupa (chrysalis), and inside this chrysalis the creature undergoes a marvellous change. It emerges from the

A pupa is another name for a chrysalis. It is an insect at the stage where it is changing from a larva to an adult insect. (The plural of pupa is 'pupae'.)

[16]

chrysalis as a moth or butterfly. The female lays its eggs and shortly dies. The eggs hatch into caterpillars.

However, only a few types of caterpillar produce a continuous thread of silk suitable for turning into a fabric. There is the tussah silkworm, which feeds on oak leaves. Its cocoons are used in some silk-producing countries to make tussah silk (wild silk). Wild silk has a rougher texture than cultivated silk, and is not so easy to dye.

The only other caterpillar commonly used for silk is called the domesticated silkworm. It eats only mulberry leaves. The first thing a silk farmer needs is a lot of mulberry trees. On the big silk farms, row upon row of trees stretch into the distance. These are carefully tended so that the leaves are in good condition for the young larvae.

When the caterpillars hatch out of the eggs they are only the size of a pinhead. The workers pluck leaves from the mulberry trees and chop them up into little pieces to feed the young caterpillars. The caterpillars never stop eating except to sleep. Those in picture [17] have increased their body weight 10,000 times in four weeks, and their length to seven centimetres (three inches). As they grow, their skin becomes too tight. It splits, and is shed. They shed their skin four times during the gorging period. Then they are ready to spin a cocoon and pupate (change into a moth).

The silkworm spins the cocoon rather like a spider spins a web. The silkworm has two silk glands which run the length of its body. These glands produce a sticky liquid substance. At

A larva is an insect at the stage of life between the egg and the pupa (chrysalis). (The plural of larva is 'larvae'.)

To 'pupate' is to become a pupa and change into an adult insect.

20

[17]

[18]

the end of each silk gland is an organ called a spinneret. The spinneret has tiny holes through which the silkworm squirts out the liquid. As the liquid comes into contact with the air, it hardens into a long silky fibre.

When it is ready to pupate, the silkworm is put into a sectioned tray like the one in picture [18]. Each silkworm has its own cell in which to spin its cocoon. Alternatively, bunches of twigs are erected above the feeding trays; the silkworm climbs up among them and pupates there. The picture on the title page shows the cocoons being collected.

First the silkworm spins a light web all round itself. Supported by this web, it then begins to spin its cocoon in one continuous thread, like cotton on a cotton reel. When it is completed, with the silkworm safe inside, the outside of the cocoon hardens.

The caterpillar takes two weeks to change into a moth. When it is ready, it spits out a substance to weaken the cocoon and breaks through into the open air.

However, if the cocoons are to be used for silk production, the moth is not allowed to emerge. If it did, it would break the long thread of silk and the cocoon would be useless. Some of the moths are allowed to emerge so that they can produce more eggs, but most of the pupae are killed. The cocoons are put into a 'stifling chamber' of hot air or steam. Then the cocoons are dried. Picture [19] shows a scene at a reeling mill in China, where tussah cocoons are being dried in the sun.

The cocoons are now ready for the next stage, which is unravelling the long thread and spinning it into yarn. I will tell you about this process in the next chapter.

Much of the work on a silk farm is still done by hand in China and other silk-producing countries. Farmers plant and tend the mulberry trees, and women pick the leaves to feed the young larvae. They can do this very quickly, stripping whole branches at a time. In other countries, such as Japan, the trees are stripped by machine.

As the trees are stripped bare of leaves, the farmer has to prune and spray them ready for next season. There must always be a good supply of leaves for the next generation of caterpillars.

[19]

4 · Making silk

When the cocoons leave the silk farm, they are taken to a silk-making factory. There they are spun into yarn.

In some countries the silk factories order cocoons direct from the silk farms. In China, where so many cocoons are produced, the farmers take their crop to a government buying station.

In China, silk spinning (called 'reeling') is still done by hand as you can see in picture [20]. It is not very pleasant work because the sticky cocoons are messy to work with. First they have to be soaked in very hot water to remove the gummy outer coating, which is called *floss*. When the gum has been softened, the workers stir the water about until they find the end of the thread, and start to wind it off.

Silk has one important difference from other natural fibres. Cotton and wool start as short, tangled fibres, which need to be spun into a continuous thread. Silk is already a continuous thread. Each cocoon gives about 280 metres of thread. However, the thread from a single cocoon is only as thick as a spider's web— much too fine for weaving into fabric. So between five and eight cocoons are used to make a single strand of silk. More cocoons are used for heavier silk. Of course, the threads are of different lengths, so as the thread from one

Spinning silk by hand

cocoon runs out, the thread of another is joined on to it. The end is twisted on, and the natural gum holds it in place.

As the silk is spun, it is wound on to reels, or into skeins.

Other silk-producing countries, such as Japan, use machinery to spin the silk. The cocoons are softened in hot water, then pass into an abrasion bath where brushes catch the ends of the threads. The brush draws the end out of the water, while the rest of the cocoon floats on down to the main reeling part of the machine. Picture [21] shows the reeling heads. Each reeling head uses the threads of seven cocoons to twist together into a strand.

[21]

When the reels are full, the silk is wound off into skeins. The skeins are packed into bales and sent to a Government testing house. Samples are taken from each bale and inspected for cleanness, strength, and a good standard of spinning. The silk is then graded, so that a price may be set for it.

Silk-making in China and Japan is an art which has been practised for centuries. Whether woven on hand looms or on mechanized looms, it is the designs on the silk that make them so special. Intricate pictures are painted or embroidered on to the silk, or woven in, using different coloured silks. Chinese and Japanese silks are famous worldwide.

China is the largest exporter of raw silk. France and Italy are the main importers of silk, which they make into fabric in highly mechanized factories like the one in picture [22]. The United States is a major importer of silk fabric.

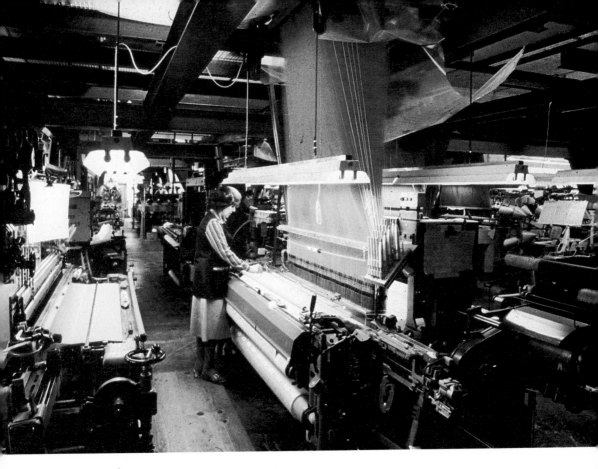

When the raw silk arrives at the factory, [22]
some is woven as it is, and some is dyed first. If
it is to be dyed first, it is washed in soapy
water to remove the gum and soften the yarn.
Yarn that is to be woven straight away is not
washed first because the gum makes it easier to
weave. The gum will be removed later by
boiling the woven fabric.

Silk yarn made from seven cocoons is very
thin, and is used only for the finest fabrics. For
heavier fabrics, several yarns are twisted
together before weaving.

The silk is now ready to be woven, dyed,

27

and finished. The whole process is similar to the one I told you about in the chapter on cotton, so I do not need to repeat it here. Coloured silks are used to make patterned fabrics, and different weaves are used in the same way as they are for cotton. Silk can also be knitted on knitting machines, and designs can be printed on to fabric after weaving or knitting.

Silk forms a very small part of the textile market—only about 0.2 per cent. It is an expensive fabric to produce because much of the work is still done by hand, and because you need so many silkworms to produce a small amount of silk. It takes 110 cocoons to make a silk tie, for example, and 630 cocoons to make a blouse!

The main producers of silk

More than half the cocoons used in the world's silk production come from China. It was there that the secret of silk-farming was first discovered more than 4,000 years ago. The Chinese kept the secret to themselves for 3,000 years.

Today, silk is made in thirty-five countries. The largest producers are China, Japan, India, Russia, and South Korea. There are also silk farms in Africa, and countries of the Middle East. Algeria is setting up a new silk-farming industry, using modern machinery to spin and weave the silk. The main silk-weaving countries, which buy the yarn to weave into fabric, are Italy, France, Switzerland, and Britain.

5 · *How cotton and silk are used*

I have already said that cotton is one of the world's main textiles, while only a small amount of silk is made. Cotton is a strong, comfortable, everyday fabric, and silk is a delicate and exotic luxury.

Look around your house and see how many cotton items you can find. You probably see clothes straight away—shirts, dresses, T-shirts, blouses, and your jeans too, because denim is a particularly strong and hard-wearing type of cotton. Other fabrics are made from cotton, such as corduroy, velvet, gabardine, and canvas. The man in picture [23] is wearing a cotton shirt and corduroy jeans. Then there are [23]

[24] all the household items, such as the sheets on your bed, table linen, towels, curtains, and the fabric of soft furnishings such as beds, easy chairs, and sofas. You can see what an important part cotton plays in our home.

Cotton is sometimes blended with man-made fibres such as polyester. This may be done to reduce the cost of the fabric, or to make it more efficient. For example, fabrics made of cotton-and-polyester do not crease so easily as pure cotton, and are easier to iron. They are useful for clothes like shirts, and household items like bed linen.

Cotton is also used in some less usual ways. Did you know, for example, that cotton rags are sometimes used to make paper for bank notes? Cotton fibre makes a strong paper which survives a lot of handling, so it is also used for some maps, legal documents, and other hard-wearing papers. The tent in picture [24] is made from waterproofed cotton. So are raincoats. Tough clothing for sports such as mountaineering may be made from a cotton fabric called Ventile. Fire-proofed cotton is used for industrial clothing, and also for firefighting garments worn by firemen.

On the other hand, if you look around your house you may not find anything made of silk. Raw silk costs twenty times as much as raw cotton, so you won't be drying the dishes with a silk towel, or sleeping between silk sheets! Silk is used mostly for high-quality clothing such as the dress worn by the model in picture [25]. It is also used for very expensive furnishings.

Other fabrics are made from silk yarn. Taffeta, satin, crepe, georgette, and organza, are all made by varying the thickness and weave of silk. Nowadays, though, these names are also used to describe man-made fibres which look similar.

You may have heard people talk about 'silk' stockings and wondered what they meant, when women's stocking and tights are made of nylon. Before man-made fibres like nylon were invented, stockings were made of real silk. Later, artificial fibres like nylon and rayon began to replace silk in popularity because they were so much cheaper and had much the same sort of feel. Silk stockings became a thing of the past. Parachutes used to be made of silk, but they are now made of nylon because it is stronger.

Nowadays, people have realised that there is nothing like the real thing. Wherever people are wealthier, and the standard of living rises, silk has become popular again. There is a demand for all the silk that the silk-producing countries can turn out. That is not only nice for those who can afford it; it is invaluable to those who earn their living by spinning or weaving this marvellous fabric.

[25]

Index

Acknowledgements for photographs: The British Van Heusen Company, no. 23; European Commission for Promotion of Silk, nos. 20, 22, 25; International Institute for Cotton, pictures on cover and page 2 and nos. 2, 3, 4, 5, 6, 7, 8, 9, 10, 11, 12, 13, 14, 15; National Cotton Council of America, no. 1; Silk Association of Great Britain, nos. 16, 18, 21; Vango (Scotland) Ltd, no. 24; Xinhua News Agency, picture on title page and nos. 17, 19.